Project Planning
for
Writing Software Manuals

Project Planning
for
Writing Software Manuals

Diane Merrall

ISBN 978-0-557-40553-4

Table of Contents

Introduction:
Project Planning Process

The intent of this manual is help both Project Managers and Technical Writers create a set of standards for writing User Manuals or other large procedural documents. The first section focuses on the content, tone, and end use of the manual—factors that are normally provided in the project scope. The second section addresses the visible features of the document style—factors that are generally not addressed in the project scope.

Why do we want to commit so much time to planning?

- To create consistency throughout the document.
  o Terminology
  o Formatting
  o Reading level
  o Instruction design
- To ensure end results meets expectations.
  o Logical sequencing
  o Focus on required details
- To establish scheduling guidelines and expectations early.
- Better project management for Project Managers.
- Better time-management for Subject Matter Experts.
- Reduce time needed for editing, rewrite, and review.

In the same way that Project Managers benefit from planning and creating project frameworks, a technical writing project benefits from establishing a document template. Once developed, the template is the framework for the documentation project. While large projects with multiple writers must work from a predetermined template to be able to combine their work, templates can also be beneficial to provide consistency across a group of short process and procedural documents.

The project **Planning** process is a critical process that reiterates the project scope and creates the document template. The resulting template provides the structure for the **Executing** phase of the project and establishes the guidelines for the **Review** phase of the project. While it may seem that a lot of time is spent in the planning phase with little actual writing accomplished, the benefit comes in greater consistency throughout the document and significantly less time needed for editing, rewrite, and review.

In any given technical writing project, writers are generally brought after the project **Initiation**. Consequently, when a writer is brought into a project, they are often dependent on documentation created in the project **Initiation** phase. The hiring firm has already made the decisions regarding business needs and stakeholder analysis, and may also have an estimate of the budget and timeline. The project documentation up to this point is generally created by the software developers, the Project Team Members, and the Project Manager.

As part of the writing project **Initiation** phase, (before the writers are hired) the Project Manager will establish the project scope. The project scope will be the source of the general topics to be included in the writing project and could also include a list of tasks to be documented. The project scope provides the writer with intended benefits and goals that software development project is designed to accomplish.

For all aspects of the project, the emphasis on the project scope is to contain or limit the size of the project. For the writing project, the scope has a direct impact on the length of the document from both a time and cost perspective, while ensuring that all essential topics are included. When a task or topic is not included in the scope, it can be considered to be implicitly excluded. And so, the project scope becomes the starting point for the Technical Writer.

For the Technical Writer, the project scope is the content agreement at the beginning of the writing project. This agreement not only establishes the content to be covered, but may also include the timelines for completion and cost to produce. While it is easy for writers to keep adding tasks or topics as the documentation is being developed, the additional content drives up the cost in terms of time to complete the project and the cost of printing the document. Consequently, the scope not only establishes the content to be included, but also creates a limit of content. Before changes or additions are made to the content established in the project scope, the impact on all costs must be considered and approved by the Project Manager.

In addition to the project scope, the Project Manager will also have documentation describing the intended process flow or workflow for the job or work that the software will support. There may be documentation that justifies the upgrade and the expected areas of improvement that the new software will provide. When additional information is available, the writer will greatly benefit by studying the work flows and justifications. The workflows and justifications are great resources for developing and setting the order to the document outline.

Planning the document template is as important as planning the content for documentation. A detailed template establishes all layout, spacing, and font parameters. The template provides consistency, which allows multiple writers to work simultaneously on various aspects of the project. The template also allows writers to review each other's work with a full understanding of the document parameters that have been established. The result is a document with consistent formatting throughout--as if the document was produced by a single writer.

Lone writers also benefit from the establishment of a document template. The document template provides a guide for handling all types of text and page layout, as well as providing consistency in spacing elements, applying numbering, and setting styles throughout the document. The lone writer always has a quick reference to verify the details, as if they are working from a checklist of formatting rules.

While the primary purpose of this manual is to address User Manuals and large documents, the concepts are also applicable for setting team standards for documentation needed to support various processes or procedures. Often, when the procedural documentation is relatively brief, the technical writing is done by a staff person without establishing standards. The result can be incomplete and vague instructions, or simply hard-to-read documents due to inconsistency in the document organization.

Using some of the layout and structural techniques in combination with creating a form for standardized information benefits both the end-user and the non-writer. A standardized format allows the end user to learn the layout and where needed information is normally located. When a document is predictable and easier to use, the information contained within the document tends to become easier to understand. The non-writer being tasked with providing documentation and can follow a prescribed format, making their job easier and they are more likely to create a user-friendly document.

While it is likely that the planning process will not cover 100% of all possible aspects of the project, it will provide a framework for handling unexpected considerations or exceptions that arise.

The Project Scope

The Project Scope section focuses on the content, tone, and end use of the manual –factors that are normally defined in the project scope. Before the writing begins, the Project Scope reviewed thoroughly to fully understand the target audience.

Scope Analysis

Before starting to write a user manual, ask:

1) *Who* will use the manual?
   Skill/Knowledge Level

2) *How* will they use the manual?
   Training, Reference, Both

3) *What* information needs to be included in the manual?
   Critical and Important Tasks, New Information

Understanding the answers to these questions helps define the end user, the degree of detail required, the content, and the organization of content. The project documentation may set the content, but it is the understanding of the knowledge and skills of the intended user that helps the writer guide the structure of the content to accommodate the user's comfort level. As a result, the end document is both user-friendly and appropriately concise.

Who will use this document?

If creating a manual as an introduction to a new software package, we may approach the project at an introductory level. When writing at an introductory level, explanations are provided for every detail and more graphics are used to step through the process. Proper detail ensures that the information can stand alone without referencing or sending the user to other sections. The only time other sections are referenced is when the information is covered in detail in a previous chapter. First-time users often find learning new software more difficult when they are frequently referred to appendices or other sections for details.

On the other hand, when writing for users at a more advanced level, we can reduce content by referencing a location where the reader can obtain more details. The assumption is that advanced users will only need the details occasionally and would prefer to see concise, condensed instruction; on the rare occasion that details are needed, an advanced user will flip to an appendix or other specified location. Also, to condense instruction, advanced users are more comfortable reading shorthand methods for describing menu options.

There are justifications for both the concise and detailed writing approaches. Certainly, the cost of writing and printing will factor in. However, it is the user's experience that is the final decision-maker. If the documentation is not usable by the end user, then it was too expensive to produce. The final document is only successful if it meets the needs of the end-user.

How will this document be used?

During the planning stage it is beneficial to establish the examples that will demonstrate the software in real tasks relevant to the environment in which this document will be used. Task examples create an opportunity to practice and review the processes, and as part of the writing process, the document will necessitate the use of appropriate examples. Knowing how the document will be used impacts which task examples are selected in several ways.

- Ensures that the screen captures make sense to the end users.
- Ensures that the written procedures parallel the training or the process.
- Ensures that the examples provided are real examples that can be followed in the corresponding environment.

Screen captures enhance the readability of the document and contribute to establishing comfort with the software procedures. By viewing examples of the screen appearance at various points, a user is able to confirm that they are interpreting the steps correctly. If the screen captures show a view different from the view they normally see, the screen shots can create confusion. Determine if there is a different screen shot for different levels of authority or different groups. If you must sign-on to the software, ensure that your profile or authorities align with standard user authorities.

Documenting software may be different from documenting how the software will be used. If the training outline, workflow outline, or process outline is available, it is important to follow the same order.

- If you are documenting the software, the documented information may be placed in an order dependent on the order of the screen menus or buttons.

- If you are documenting software for the purpose of teaching a process, the documented information is placed in the same order as the workflow or process.

It is beneficial to establish the specific examples or datasets that need to be used as examples so that the screen captures provide an accurate example. If possible, the examples selected should be real details that the user will find logical. For example, if you are displaying how a search function works, find out how users will typically use this function. If the examples are illogical or arbitrary, choices, they become counterproductive.

It is possible that there is a software training database or environment that the users will use for practice, or there may be specific circumstances that will need to be addressed for practice. Establishing real test, training and practice parameters can strengthen the use and quality of the manual.

What will be included in this document?

The Project Scope is interdependent on the Topics selected, the Outline, and the Length and the Cost of the document. The Scope establishes the Topics to be included in the manual. The Outline sets order to the Topics and determines the Length of the document. The Length of the document has significant impact on the cost from the perspective of both "time to produce" and "cost to print". As the tasks are selected and the outline created, the length of the document and the size of the project become apparent. This is significant for both determining the cost of the project and establishing a timeline for completion of the project.

The project scope may come from a variety of sources. The project scope may already be written as part of the software development project plan, software installation project plan, or the software purchase project plan. If the software will support a process, it is likely that process was documented prior to developing the software; if so, the process document may work in conjunction with the scope for the writing project. If the writing project is for a software development company, there may be a sales or marketing plan; if so, that plan may also contribute to the project scope.

The outline is the child of the scope in the planning process. The outline provides details while the scope provides general concepts. The outline becomes the structure that displays the order and information of the document. From the outline, the project can be divided into writing segments for individual team members, and required heading styles can be determined.

At the outline level, the Project Manager can ensure that the project cost and the project scope are in balance. If the estimated cost is outside the projected budget, the budget is brought back in line by reducing the length of the document, pulling items out of the outline, and possibly removing one or more topics. Any change to one aspect impacts all other aspects.

Wrapping Up - Scope Analysis

1. Who will use the document?
   a. Are they advanced users?
      i. Concise
      ii. Few screenshots
      iii. Shortcut language
      iv. Reference the details – create appendices
   b. New Users
      i   Provide detail – menus and buttons
      ii Screenshots as frequently as possible
      iii Full sentence descriptions
      iv Avoid references to other sections
2. How will the document be used
   a. User Training/User Manual
   b. Technical Guide
   c. Administrator Guide
   d. Quick Reference Guide
   e. Applicable detail for each type:
3. Examples
   a. Screenshots needed
   b. Datasets
   c. Training environment
   d. Terminology
4. What will be included
   a. Topics
   b. Sub-list of tasks for each topic

The Outline

Task List

The outline begins with a list of topics as specified in the scope. The topics selected are determined from an overview level and should coincide with the primary reasons for the software package. For example, if the software being documented is Microsoft Outlook, the two primary objectives for selecting this software may be to handle email and schedule events. Therefore, the primary topics for the outline are:

    I. email
    II. calendar

Following each primary topic will be a list of subtopics. Subtopics will list the general tasks for the end user. If end-user of this manual is a technician who installs and maintains the software, the subheadings are directed toward the configuration of E-Mail accounts. If end-user of this manual is someone who will send and receive email, our subheadings will be focused on the processes necessary to handle incoming and outgoing E-Mail, such as creating, sending, and adding attachments. The outline for the may include the following subtopic tasks:

    I. E-Mail
       a. Create
       b. Sending
       c. Attachments

The list of topics and subtopics may also be influenced by the familiarity with past versions of the software. If the purpose of the manual is to provide instructions for a new release of familiar software, the primary topics list will be directed toward the most frequently used tasks. It may also be headed up by a "Changes" section. The details that users should already be familiar with are excluded from the main content, but may be included in a Quick

Reference Guide in the appendix. Examples the details that a user should already be familiar with may include a table of shortcut buttons or a table of keystroke shortcuts.

If the purpose of the manual is to introduce a new software package, the primary topics may be a list of the building blocks necessary to do be able to introduce basic or everyday tasks in the new software. The "new user" flow may require that subtopic tasks include the introduction of each screen, menu, and button along with a list of essential and commonly used terms, keystrokes, and shortcuts.

Order of Tasks

After determining which tasks need to be included, the tasks are grouped into three groups – the core tasks, essential work-flow tasks, and high-level skilled tasks.

The first level tasks are the core tasks. These are tasks that need to be understood before a subsequent task can be introduced. As an example, when you work in most software packages, the text needs to be selected before you can take action on the text. Consequently, in terms of the order of topics, the section on 'how to select text' would come before the section on 'take action on the text'.

The second level tasks are the tasks related daily work-flow tasks. This is the hardest group to establish order, unless detailed work-flow documentation exists. These are the tasks that are performed regularly and are essential tasks in the completion of a job function. For example, a Help Desk may enter tickets for calls, or an accountant may enter billing information multiple times.

The high level tasks are procedures that may be used infrequently, may apply to few users, or are more difficult tasks.

The resulting three groups provide sections for the manual. The next step is to establish sequence within each of the sections. Within each task group, there is an order that is determined by either following the work-flow process or the software process. Following the work-flow process simply means determining the order starting from step one of a process or creation of an object, stepping through intermediate changes, and finally, the closure or completion.

Once the tasks are grouped and sequential considerations have been addressed, it is possible that there will still be multiple items that need to be ordered. There may be several core, workflow, or high-level topics that are competing with each other for "next in line". To help sort out topics that appear to be similar in importance, evaluate the frequency of use and the ease of use. Sorting by frequency of use brings topics that are more in demand to front, with the first group being everyday procedures and all non-frequent procedures following in the second group.

Within the high frequency topics, establish an order from the simple to difficult. By selecting topics that will be used daily, you are addressing the importance of the topic in relationship to the users work. By addressing simple topics before more challenging topics, you help build user confidence before challenging them.

Within the remaining topics that may have a low frequency of use, consider task grouping. Often, you will find that specific project requires a different set of skills. Find the procedures that apply to specific processes, job descriptions or projects. Task grouping often comes into play with higher-level topics.

For example, a word processing task group example may be "Creating Long Documents", which would include the sub-topics of "Headings", "Table of Contents", "Indexing" and "Headers and Footers". Likewise, outlining "Mail Merge" as a task would include the subtopics of "Creating labels", "Letters and Envelopes", and "Data sources". In a particular situation, The Mail Merge group of tasks may be considered more commonly used and so would be placed before the Long Document group. Making the determination of order at this point would require understanding how the document will be used.

Also consider grouping tasks based on the end result. Group tasks that work together to create a newsletter, a report, or other type of project. If the manual is going to be used in a training environment, the task grouping will also consider the amount of material that will be covered in a given period of time.

As you reach the task grouping topics, it is generally safe to say that the user level is at least at a moderate level. It is generally acceptable by this point to refer users to early chapters for basic skills.

As the tasks are ordered, it will be discovered that not all of the listed ordering options are needed. There will be times when the workflow process is all that is needed to order the process. The more tools that that the writer has to order the information, the better job they will do at creating an outline that is comprehensive and still has a logical flow.

By the time the outline is complete, the tasks should be in order of a natural flow of topics. The most basic tasks or the building block tasks are introduced first. In terms of introducing new screen elements, the progress is to move from left to right and top to bottom. The subsequent tasks progress to more difficult tasks. The outline is the optimum view of the document layout. If the sequence doesn't look right in the outline, it will not flow right in the manual. This is a significant review point for both the writer(s) and Project Manager.

As the list of tasks is created, the list of necessary prerequisite skills may also become apparent. If the prerequisite skills are not included in the project scope, the information regarding the prerequisite skills is put into the introduction. Once the introductory sections are written, the project manager will make a decision on how to handle the prerequisite skill information. They may choose to add the prerequisite skill information in an appendix, add it in a chapter or let the Introduction state the prerequisite information.

As the Outline is completed, the workload for each section also becomes apparent. The Project Manager or the Writing Team Lead is able to breakout sections and assign to individual writers.

Wrapping Up - Outline

1. Create Task List
2. Order Task list
   a. Essential basic skills
   b. Frequently used tasks to less frequently used tasks
   c. Easier tasks to more challenging tasks
   d. Group task sets toward end projects

The Introduction

Each chapter and section begins with an Introduction. The Introduction provides a brief general introduction to the topic and includes the goals of the section, potential work applications, and prerequisite skills and knowledge when appropriate. If this software is used as part of a larger process or workflow, the workflow information may also be included in the Introduction.

For the user, the Introduction states the purpose and use of the information contained in the section or chapter. The Introduction gives the user an idea of why they want to learn this particular topic, how it will enhance their work, or when they will use the information.

The Introduction to each chapter or section, should also address the level of knowledge or skills that the user will need to successfully complete or process the information. The level of knowledge advises the user regarding the skills and knowledge needed to complete this section or chapter. If additional preparation is needed prior to reading or using the information, it is beneficial to provide a reference to the preparatory material. The reference may be in the form of a web address or may refer the user to an earlier chapter in the manual. References may also include the suggestion of training options.

The Introduction may include terms and screen captures to establish a common language. Where appropriate, the screen captures introducing the upcoming topic are marked up with call-outs to ensure a common naming system for various toolbars and menu bars. When the software is new or graphics and buttons have changed since the last version, a table of buttons can help smooth the transition to the new software.

Consider the following items for Introduction:

1. Screen shot of the initial working screen.
   a. Identify toolbars
   b. Identify workspaces
   c. Table of buttons
   d. Button list

    e. Button picture/screen capture

    f. Button functions

2. Vocabulary

    a. Changes in vocabulary from last version

    b. New concepts

When the software is used as part of a workflow, it is beneficial to obtain the workflow information to include in the Introduction. There may be specific automated processes that triggered by the software inputs, or there may be text fields that will be captured and provided in reports to other team members, management or other interested parties. When input fields can result in expanded communications, it is advisable that the manuals provide the appropriate workflow information.

The Introduction is the appropriate place to include a paragraph related to specific use requirements or limitations, and company policies or issues that that user needs to know. If the company is following established national, governmental or global standards, providing a web site link to standards compliance information will allow the user to access the standards details and will reinforce the significance of the procedure documented.

In addition to providing the reader with a set of expectations, the Introduction also serves as an organizational tool for the writer and a project management tool for the Project Manager. As the Introduction is written before the content, the project managers have an opportunity to address scope issues early. This includes making adjustments to address prerequisite knowledge.

Wrapping Up – The Introduction

1. Topic discussion
   a. Significance of topic
   b. Application of information
   c. Prerequisite knowledge and skills
   d. Common Language
2. Screen Shots
   a. Identify toolbars
   b. Identify buttons and menus
3. Vocabulary
4. Company Standards
5. Workflow or automated processes

The Summary

The information included in a summary section is strongly influenced by how the manual will be used. When the manual will be used in a classroom, a Summary may be created as a review of the topic. Training manuals generally include additional practice examples at the end of each chapter. Training manuals also call for review of terms and procedures to reinforce the chapter as a learning experience.

In a training guide, the Summary should tell the user what they just learned. In both user manuals and training manuals, the summary may provide guidance about what the user is now prepared for and where they might apply their new knowledge. In addition, the Summary may guide the user on additional related topics, both within the manual and beyond the manual.

When available, a training manual will list additional reference materials or websites. The additional references may provide white papers that demonstrate additional applications to reinforce the knowledge. Additional references may also provide informational opportunities on how to expand knowledge beyond the scope of the manual.

Both user manuals and training manuals can benefit from a quick reference guide at the end of the Summary. The quick reference guide is a brief document that gives condensed outline of the procedure for future reference. When it is placed at the end of the chapter, or at the end of the manual, the quick reference guide is either a tear-out page or a page that can be copied and posted in a convenient location.

The quick reference guide is written with only the most basic information and is often set up in a 2-column format. The result is the appearance of a cram sheet, where space is a premium and key words are all that is needed. The instruction provided is only the most basic information needed to get from point A to point B. The accompanying screen captures are reduced in size to fit into columns.

The value of the quick reference guide for the user is to act as a reminder or a review. The quick reference guide is not a replacement for training or a manual, but it does provide a way to quickly look up a forgotten step.

Creating practices exercises requires knowledge of the workflow for the environment or situation that the user will experience. In the same way that the examples used to create screen captures must portray real situations, the practice exercises at the end of each chapter must also represent a potential situation.

Whether you are practicing making a reservation, creating a call ticket, or setting up a mail merge, the user needs to see the practice example as being something they could actually do on their job. Expecting a user to make the leap from an unreal example to a real application can leave a user feeling as if the practice exercise is not worth their time.

When available, a training manual will list additional reference materials or websites. The additional references may provide white papers that demonstrate additional applications to reinforce the knowledge. Additional references may also provide informational opportunities on how to expand knowledge beyond the scope of the manual.

For all the ideas that may appear in a Summary, this is the one section that may not appear at all. The use of the Summary section seems to be primarily related to the use of the manual in training or education environments. The more formal the training environment, the more likely the Summary will contain multiple facets. The less formal training environment or if the manual is intended to be a user manual, the less likely there will be a strong focus on a Summary. When there is not a Summary, the quick reference guide may be created as a stand-alone document.

Wrapping Up – The Summary

1. Summary of Topic

2. Quick Reference Guide

3. Practice Exercises

4. Additional References or websites

Project Organization

The Project Organization includes the financial aspects of the project which includes both the time to write and cost to print the project deliverables. The Project Organization also includes designating storage areas for graphics, work-in-process, and completed work.

Estimates for Deliverables

The Project Manager may have already established timelines and budgets for writing, review, and printing before a writer is assigned a project. In fact, when the writer is brought in, the writer will often receive the project estimates as part of the writing project description. However, the writer needs to aware of "normal" timelines and be able to determine when the Project Manager time-lines are appropriate or unrealistic.

The Project Manager will also have a defined list of deliverables. The deliverables would include the User Manual, but may also include an Index, Glossary and/or Quick Reference Guides.

To determine a rough estimate of time, the average writer will need 3 – 4 hours per page. This includes research, interviewing subject matter experts, writing, developing graphics and screen captures, and editing. If indexes and glossaries are required, add another 1 -2 hours per page.

Organization for Project Work In Progress

It could be a great time loss to a project if a team member abruptly becomes unavailable or does not have a backup copy of their work. To ensure that work-in-progress is accessible and properly backed up, the writer(s) need to establish a known shared location. All work-in-progress needs to be stored on a server with access for the Project Manager and the writing team.

If something happens to a team member or their equipment, the rest of the team needs to have the ability to recover the work. This is just as important for a lone writer. The Project Manager could be placed in a situation where the project needs to be started from the very beginning if the writer suddenly becomes unavailable. In addition, with all work in a central location, the Team Leaders and Project Managers are in a position to periodically review work to ensure that the project is on track.

Where all members of the team are working on-site, it is practical for each member to be working on documents stored on a server. In a situation where the team members are working remotely, they may be required to upload or email the most recent version of their respective sections at least once per day.

The Project Scope Summary

At the completion of the outline and the Introductions and Summary's, the project should be well-defined. The project Scope may have been adjusted as the outline and introductory sections were developed, or the outline may have been adjusted to keep within project scope parameters. Everyone has reviewed the outline and is in agreement that the outline represents the best flow for the material.

At this point, there should be an agreement on the following areas:

1. Scope Statement
    o   Document outline ordered
    o   Topics
    o   Subtopics
    o   examples
    o   The Introduction – rough draft
    o   The Summary – rough draft

2. Identifying deliverables
    o   Manual length
    o   Degree of detail
    o   Screens captures
    o   References
    o   Supporting documents
    o   Manual
    o   Quick Guides

- o Index
- o Glossary

3. Estimated Timelines
   - o Writing
   - o Review
   - o Printing

4. Estimated Costs

5. Work-In-Progress maintenance

With the agreement on the Scope complete, you are ready to develop the parameters for the document template and style guide. The Outline will provide the framework for the applying the Style Guide. The Outline brings the format requirements to the front by showing the number of heading levels that may be needed. While the style guide can be developed concurrent with the development of the Outline, the Outline is the tool that puts the style guide to work, ensuring that the style guide is providing adequate number of heading styles.

Developing Templates & Style Guides

Developing a template and style guide produces a set of document standards promoting consistency in the document. It doesn't matter whether the document will be written by a single writer or by a team of writers; without a template and style guide, it is difficult to maintain consistency. Without a style guide, both writers and editors waste time comparing sections to be sure that each section is handled in a consistent fashion.

The Project Manager may contribute certain style parameters for purpose of consistency or cost management. If the Project Manager or their company has a series of manuals, they may have style guides to be applied or require that previous manuals are used as a standard. In addition, the Project Manager is concerned with cost. As the styles used in a manual may impact document length and ultimately document cost, it is reasonable for the Project Manager to have final approval on the document template and styles. Consequently, the template and style guides need to be reviewed and confirmed before the writing begins. As part of the template, we will develop style guides for the following areas: Page Layout, Graphics and Lists.

Page Layout

The page layout parameters establish the landscape for the document. Each contributes to the readability, ease of navigation and the overall flow of the document. The page layout parameters include:

- Margins & Indentations
- Paragraph Settings
- Headers & Footers
- Starting new pages

Margins

The margins required are determined by how the manual will be bound. To accommodate most binding types, the margins will be set to "mirrored', with additional inside margins. For long documents over 100 pages, the recommended binding options are Three-Ring binder or Perfect binding. For shorter documents, additional binding options are Saddle Stitch or Spiral binding.

Saddle Stitch is the only option that does not require additional inside margin considerations. All other binding options require the additional inside space provided by mirrored margins. The mirrored margin creates an additional ¼-inch margin on the inside margin, allowing extra space required for binding.

To picture the additional space needed for Perfect binding, look at a paperback book. As the book is opened, it does not lay flat and the inside edge of the page is tucked into the binding. Three-ring binders and spiral-bound documents will use the extra margin allowance for space to drill holes and Perfect binding will need extra margin allowance to accommodate trim and the area that is glued to the binding.

This means that as you develop your document, all right-hand pages will have the inside margin on the left and all left-hand pages will have the inside margin on the right.

Indentations

The width of the column of text can significantly impact the readability of the text. Generally, narrow columns of text improve the ease of reading. The goal is to set the width so that a reader's eye has minimum side-to-side movement. However, if the width of the column is so narrow that words frequently break into hyphenated syllables, the column is too narrow.

At a minimum, all text is set with a one-inch indent from the left margin. This means that if the left margin of the page is 1 inch, then the text will begin at 2 inches.

In a document with indented text, the headings are aligned at the left margin- in other words, no Left indentation. When the headings align with the left margin and the text or content is indented one inch, the headings are easy to see as you scan the document. The position of graphics and tables should follow the indentations of the text which allows the headings to stand out.

Guidelines for Margins and Indentations

1. Set page margins to Mirrored margins. This sets wider margins on the inside and standard margins on the outside edges.

2. Left align Headings to the standard page margin (1 to 1.25 inches from page edge).

3. Left align the text to the standard page plus 1 inch (2 to 2.25 inches from page edge).

4. Size screen captures and tables to fit within the margins of the text area.

Paragraph Settings

The paragraph settings contribute to the balance of white space and facilitate easier reading. On a page where there is a large proportion of text with minimal graphics, creating additional white space through paragraph spacing is more relaxing to the eye. On pages where there are more lists or graphics, the appropriate use of white space guides the reader from one step to the next step or one section of text to the next section of text.

The paragraph settings determined in the template include the line spacing, the first line indentation style, and the space between paragraphs. Although it appears that this should be a simple list, a typical User Manual has a variety of paragraph types to be considered by the writer. Paragraph settings need to be determined for the following types:

- Standard paragraphs
- Numbered lists – usually instructions
- Bulleted lists (unordered lists)
  - Long item lists (more than 2 – 3 words)
  - Short item lists (less than 3 words)

Standard Paragraph Styles

The optimum readability of a document is primarily facilitated by using language and sentence structure geared toward the target audience. But the document layout also has a significant impact on the readability. A layout with 30% white space allows a reader to skim through the content to pertinent information.

If a reader is skimming through material, they may choose to skip to the next paragraph after reading the first sentence. To accommodate the reader's ability to navigate easily between paragraphs and through various paragraph types, each paragraph is distinguished with extra space. For standard paragraph styles, the distinction may be provided by an extra line between paragraphs or by indenting the first line of a paragraph.

When extra space or an extra line is added between paragraphs, there are two methods of creating the space between paragraphs. When the extra space between paragraphs is equal to a line height (or double-space), some writers may create this extra space by hitting the enter key one extra time. The second option (and preferred option) is to set the extra space between paragraphs in the "Paragraph Settings" option in the desktop publisher software.

It is preferred that space between paragraphs is set in the desktop publishing software paragraph settings. Using the settings in the software ensures that the height of the extra line is consistent and ensures that future updates or changes are consistent as well.

When the document is laid out in a single-spaced format, adding space between the paragraphs is necessary to create white space and assist the user in distinguishing the beginning of a new paragraph.

The additional space added between paragraphs should be the minimum of 6 points or approximately half the number of points used within the text lines of the paragraph. The maximum additional space is equal to the number of points for a given line. Typically, with a 10-point font, the line height is 12 points; the extra 2 points ensures that in single-spaced layout, the letters for each line are not on top of each other.

When extra space between paragraphs sets a paragraph apart, you may see that the first line of the paragraph is not indented. Typically, this will be used in communications where the paragraph line spacing is single-spaced, with double-space between the paragraphs.

When the document is laid out with 1 ½ line-spacing or double-spacing, extra space between paragraphs would be excessive. Consequently, 1 ½ line-spaced or double-spaced documents must have an indentation to distinguish each paragraph. Again, it is preferred that the indent is set in the desktop publishing software paragraph settings so that document layout changes can be applied at a later date.

Guidelines for Standard Paragraph Styles

1. Single space paragraphs
   a. Minimum of 1 ½ lines between paragraphs.
   b. Maximum double-space between paragraphs.
   c. Indent first line optional with extra space between paragraphs.
2. 1 ½ or double spaced paragraphs.
   a. First line indentation is necessary.
   b. Not necessary to add extra space between paragraphs.

Paragraph Styles for Ordered Lists

A User Manual involves often requires the use of lists to create the steps in a procedure, but may also require the use of bulleted lists, or outlines. The numbered steps in a procedure and the items in an outline are identified by the desktop publishing software as paragraphs. Desktop publishing software defines a paragraph as any text followed by the press of the Enter key. Consequently, the paragraph styles that are set for standard paragraphs will also apply to list style paragraphs. When standard paragraphs in the document have been set to "6 points to follow", the process will be automatically applied to both lists and standard paragraphs by the desktop publishing software.

As the goal of writing a User Manual is to provide easy-to-understand instructions, it would follow that we want the instructions to be easy to read. User Manuals are read in a similar fashion as a cook book. Generally, the reader of a User Manual moves back and forth between the instructions and the computer screen. Appropriate white space around each instruction assists the reader in relocating each step as they look back to the manual to move through the process. The appropriate white-space-to-text balance allows the reader to distinguish each step within the process. Also, when the line spacing is distinctive from standard paragraphs, it can help the reader see the sequence of steps as a cohesive process.

Each step in the instruction is considered to be a single paragraph, so to create balanced white space around instruction, both line spacing and indentation are addressed. Generally, as soon as you apply numbering to an item, the desktop publishing software will indent the numbered items by ½ inch. Recall that the text indentation is set to 1 inch, so the total indentation for numbered items becomes 1 ½ inches.

The paragraph setting for the instruction is single-spaced. Setting the Special Indentation to "Hanging" will indent all text by ¼ inch. The result will be as follows:

1. The number is indented by ½ inch and the text has a "hanging indent of ¼ inch.

2. Each subsequent instruction will display the number easily.

3. In addition, the instruction set is distinguished from the rest of the text.

Allowing an additional 6-point line space between each step (paragraph) in the instructions creates adequate white space to allow the eye to identify each step. Notice that the text within the instruction is single-spaced, but the space between the instructions is 1 ½ spaces. The 1 ½ lines spacing is a consequence of the extra 6-point spacing that follows each "paragraph". This is different from setting the line-spacing to 1 ½ line; setting 1 ½ line spacing would have set the spacing within the instruction to 1 ½ as well as between the instruction.

Paragraph Styles for Unordered Lists

Generally unordered lists or bulleted lists that consist of a list of brief items (items that consist of 1 – 3 words) have sufficient white space around the list and do not need additional white space between the line items. Consequently, for lists of brief items, the spacing is set to single-space with no additional space between the line-item paragraphs.

- List Item blue
- List Item green
- List item red

Adding extra space within the above list of brief items would make the list appear to be fragmented or lacking in cohesion. In addition, excess white space can appear to be wasted space and detract from the list.

There may be cases where an exception would need to be applied, such as when several lists are related, or the list was inconsistent in the length of each item. This would be addressed in the **Review** process when the document is in a rough draft form.

Guidelines for List Paragraph Styles

1. The numbers or bullets should start at the same vertical point that a new paragraph would start.

2. Text is indented ¼ inch from the number or bullet. Use "hanging" paragraph styles for proper indentation.

3. List paragraphs are single-spaced text within the item paragraph.

4. List paragraphs spacing between the paragraphs is dependent on the length of the line item:
   a. If the line items are more than one line long, there should be at least an extra 6 points between items.
   b. If the line items are all single-line, short items, do not include extra space between lines.

Headers and Footers

Headers and footers provide document navigation guidance for both the reader and the writer of a document. At a minimum, the pages need to be numbered, with the numbers appearing in the footer and positioned in the center or to the outside edge. It is also beneficial to include chapter or topic information in the header. If the chapter information is included in the footer, the chapter information is on the inside and the page number is on the outside edge.

There are 3 positions of the header and footer to consider, but not all positions carry the same ability to draw attention, nor should all positions be used at the same time. The center is only used if there if there is only one item, such as page number or book name. The inside edge is only used if there is additional information on the outside edge. The outside edge may be used in either situation; when there is information on the inside edge or when there is no additional information. In addition to positions within a header and footer, odd and even pages can be handled with different headers or footers.

The inside edge of the header and footer is for less important or less used information. The center or outside edges are positions of more dominance because it is more readily viewed. Consequently, the information that belongs on the outside edge of either the header or the footer is the information that the reader will most often need, such as page numbers.

For simplicity in managing Header and Footer changes between sections, consider a system where page numbers and items that apply to the entire book or manual are in the footer. The chapter names and other items that will change with each section, go in the header. This allows the footer to stay the same across sections (the footers are linked across sections), but the header changes within each section (the headers are not linked across sections).

Guidelines for Headers and Footers

1. Use of even and odd page headers or footers
2. Header:
   a. Name of the Manual
      i. Outside edge
      ii. Centered– only if no other information is included
   b. Name/Number of Chapter
      i. Outside edge
      ii. Centered– only if no other information is included
3. Footer:
   a. Always Page Number
      i. Centered – only if no other information is included
      ii. Outside edge
4. Manual Name
   a. Inside edge
5. Name/Number of Chapter
   a. Inside edge

Starting New Pages

The start of a new page signals a significant change in topic. Placing a heading at the top of a new page sends a stronger signal than the same header in the middle of a page. So, to bring attention to the beginning of a new chapter, the first page of a new chapter always begins on a new page. New chapters also start on a right hand-page. This may leave a blank page or a page with little information on the page before a new chapter, but it is necessary to clearly distinguish the new chapter.

Topic headings are also excellent points for starting new pages, although it may not be practical to start all new topics on a new page. If the template does not require a new page for a new topic, then it must be stated that new topics should only begin in the top half of a page. Topics that begin in the bottom half of the page appear to be less significant, regardless of the Heading style used. In consideration of the length of the document, specify the accepted page position where a new topic may be introduced. Acceptable positions for starting a new chapter are the top quarter, third, or half of a page. Topics should not be started in the bottom half of a page.

By starting a topic on the top of a page, we accomplish two things. First, as already discussed, it draws visual attention to the beginning of a new topic. Second, starting a topic on a new page will reduce the possibility of leaving fragments of instructions on the first and last pages of the instruction set.

Consider a 2-page that starts in the middle of the page. It will take 3 pages to cover the topic; a half page to begin with, a full second page, and the final half page to complete the topic. The result will be an instruction set with a fragment on the first page and a fragment on the last page. The fragmented lines are lines left as orphan lines, while the rest of the instruction is on another page. Orphaned lines can be easily overlooked, making the instruction appear incomplete or just more difficult to follow.

Fragmented sections or orphaned lines can appear at the end of the instruction as well as at the beginning, even when you start a new topic at the top of a page. So, when evaluating how the instructions are spread over pages, also look at where the instructions end. If the final page of the instructions contains only a few lines, look for opportunities to compress or expand the graphics or information so that the orphaned lines at the end are not forgotten. In the same way that instructions should not begin with a few orphaned lines, we also don't want the instructions to end with orphaned lines.

Guidelines for Starting New Pages:

1. New Chapters
   a. New Page – always
   b. Right-hand Page – always
2. New Topics
   a. Always new page if topic fits on one page
   b. New Page if:
      i. New topic would begin in bottom ½ of page
      ii. New topic leaves less than a 1/4 pages on last page
   c. Follows old topic on same page if:
      i. New topic begins in middle of page
      ii. There is a desire minimize pages

Font Selections – Text

The base font is the core of all the styles that will be applied in any given manual. Any style deviations will be dependent on the base font. Consider base font to be the tool that conveys the message–Professional and Polished, but understated. Select a font that is easy to read, such as Times New Roman, Garamond or Verdana. Avoid fonts that have their own message, such as 𝕁𝕠𝕜𝕖𝕣𝕞𝕒𝕟 or Comic Sans.

Before addressing fonts, it is beneficial to understand the font descriptors, "type" and "style". Both the font type and the font style are necessary to completely describe the font used in each situation and application.

Type – the font design described by the name of the font. For example, Times New Roman is a font Type. It does not tell you anything about size, emphasis, or color. The font type is incomplete without more information, but you do know the shape of the letters and whether the font is serif or sans serif. The serif is the extra stroke added to fonts such as Times New Roman, Cambria, or Garamond. A sans serif font is without the extra stroke. Sans serif fonts include Verdana, Calibri, or Tahoma.

Style – the variation of the font design. You may italicize the font, bold the font, change the size, or change the color.

When establishing the template or style sheet, the normal or standard font type and size needs to be determined. The default font type is the font used for "Normal" text – normal text is the text used in paragraphs (and instructions!). The normal font sets the basis for all other emphasis and heading fonts. Generally, the Heading fonts will be the same font type as the normal font type, but with higher point values and added emphasis through bolding, italicizing, or underlining.

In the rare situation that you use multiple font types, the font types should be limited to 2 types with one font used consistently for Headers and the other for text used within the body of the document. The various heading levels and emphasis styles should be achieved through font style changes, not type changes.

The key determination for font selection is readability. There was a time when sans serif fonts were determined to be the most readable on-screen and serif fonts were selected for print, but attitudes have changed in recent years, and writers are no longer constrained by the rules regarding serif. Regardless of the font selected, consider readability before cuteness or uniqueness; avoid fonts that are compressed, fancy, or bold. Remember, professional and polished, but understated.

Heading Levels

The quality and consistency of heading levels is important for both the development of the long-document features and the readability of the document. The consistency in the use of heading levels will allow the desktop publishing software to produce a quality table of contents. In addition, updating a table of contents is accomplished with a few clicks of the mouse when table of contents is based on the consistent application of headings.

Using the aid of the developed document outline, you can see the heading levels that are needed to assign the correct or appropriate heading levels. The chapter subjects define the major topics in the outline. Heading 1, the most emphatic heading is applied to the chapter names. The topics within a chapter are assigned Heading 2. Heading 2 fonts are the same font type as Heading 1, but 1-2 points smaller and possibly less emphasis in color or boldness. The font for sub-topic headers would be smaller and less emphatic to give the reader a clear understanding that they are still under the same topic, but have changed sub-topics. The selected heading font styles become part of the outline to ensure that the styles are used consistently throughout the document.

It should also be noted that there are special heading styles for titles and captions. To allow titles and captions to be included correctly in the table of contents, they need to be assigned the proper style.

Font Emphasis

The purpose for font emphasis is to draw attention to specific words. This is not accomplished by using heading styles! *Warning* – when you use heading styles to apply a style to text, the text will end up in the table of contents. Best practices create emphasis through bolding, italicizing, underlining, capitalization, or changing the color. Also, consider using text boxes or outlined tables to draw attention to information.

The primary guidelines for applying font emphasis are:

- Apply emphasis consistently.
- Use emphasis sparingly.

Creating emphasis on specific words and phrases helps the reader identify key items. As the template is developed, decide what an emphasized word means and what type of emphasis should be applied for specific situations. If key vocabulary words are bolded, then each time a reader sees a bolded word, they will identify that word as a key vocabulary word. If menu items and key vocabulary words are both bolded, the message created by applying bolding becomes less clear.

Further, once you determine what a bolded word means, be sure to apply bolding each time that situation arises. If bolded words are menu items in the middle of instructions, be sure that all menu items in the middle of instructions are bolded. If key vocabulary words are italicized and you have a glossary, be sure that all words found in the glossary are italicized when they are found in the text AND all italicized words in the text can be found in the glossary.

Creating emphasis with bold, italics, and capitalization often transfers well when the document is printed. Consequently, bold, italics and capitalization are the preferred methods for emphasis and why they are used often. Changing font size for emphasis is generally limited to headers only; in the middle of text changing font size is avoided because it not only changes the size of the selected letters; it also changes the height of the entire line of text. The end result is that one line looks unusually larger than the rest.

Using color can help draw attention to certain text objects, but this only works if the document is printed in color. If the document is going to be used on-line, the color may help words or headlines stand out, but too much color can create distraction. If you use a color to draw attention, there are two rules to follow:

1. Plan to be consistent with that color.
2. Test print in black and white.

Be sure that when you print in black and white, you can still identify the call-out. As an example, yellow will print as a very light color, so may not be visible and may not have sufficient impact in a black and white document. Red is a good color for call-out box outlines and arrows; it has enough contrast in a black-and-white document to be visible. While the red color is not distinguishable the black-and-white print-out, the callout box or arrows will still draw attention. When the red object is viewed in a color document, it draws immediate attention.

The use of emphasis should be limited to less than 5% of the text. It is intended for emphasis on a word or brief phrase – nothing more. When more than 5% of the text is emphasized, the emphasis becomes confusing and convoluted. The purpose of emphasis is to draw the reader's eye through the steps and to key words. If colors, boldness, and capitalized words are used liberally, nothing stands out; it becomes loud and noisy.

When a warning or additional information is called for, it is better to set the text off in an "information" box or a "tips" box. In the template, establish the width of the box, the line-weight, graphic to be included, font of the message and indentations within the box. Consider the impact of the following three options for warnings or additional information:

1. Italicized paragraph:

   The primary purpose for changing the font style for emphasis is to assist the reader in moving through the instructions or text. The style needs to be esthetically pleasing, but the esthetic aspect is a secondary objective, not a primary purpose.

2. Highlighted textbox:

   > The primary purpose for changing the font style for emphasis is to assist the reader in moving through the instructions or text. The style needs to be esthetically pleasing, but the esthetic aspect is a secondary objective, not a primary purpose.

3. Highlighted NOTE:

   NOTE: The primary purpose for changing the font style for emphasis is to assist the reader in moving through the instructions or text. The style needs to be esthetically pleasing, but the esthetic aspect is a secondary objective, not a primary purpose.

In the previous examples, the italicized paragraph is more difficult to read and does not stand out as clearly as the boxed text. The boxed text is the same font style and type as the normal font, so it is easy to read, but the box brings attention to the text. While the red will not come through in a black and white print, the box will draw attention. The third option, using a capitalized, red NOTE: to draw attention to the text works well for on-line, but in a black-and-white printed document, only the capitalization creates attention.

Using a boxed text is a good way to draw attention to more important messages or warning messages. Using the "NOTE:" option may be more suitable for when providing additional information after an instruction set. The NOTE: option lets the reader know that there is additional information, but that this information is not one of the steps in the procedure.

Remember that the primary purpose for changing the font style for emphasis is to assist the reader in navigating through the instructions or text. The style needs to be esthetically pleasing, but the esthetic aspect is a secondary objective, not a primary purpose. Guide the reader, but be careful not to create distractions with excessive or difficult to read emphasis.

As a final note to the use of font emphasis, do not try to use heading styles as a shortcut to applying emphasis. Heading styles are only for use in headings and are not to be used to create emphasis in normal text. When Heading styles are applied, they create "bookmarks", which will be used by desktop publishing software to create hyperlinks and tables of contents. When Heading styles are applied to text in the middle of a paragraph or instructions, you will find those items appearing unintentionally in your table of contents or other reference tables.

Fonts for Instructions

In the process of developing a User Manual, you will write instructions that will not follow conventional sentence structure. The use of font emphasis on key words helps the reader understand the information you are trying to convey. For example, the statement,

"Click on Insert Diagram", needs a little assistance to fully understand the meaning. As the sentence is written now, you could be clicking on a button called "Insert Diagram", the menu option called "Insert" or "Insert Diagram", or a sequence of menu selections – "first click on Insert, then click on Diagram".

Prior to writing, develop a plan that conveys the format for instructions. Designate the sentence structure to be used to describe how or where to click on menu items, toolbar buttons, or keystrokes. Also, determine how to designate a series of menu selections. Sometimes, the specific steps for menu options are written out as shown below.

Menu options written out:

1. Click on **Insert** on the standard menu bar.

2. Click on the **Diagram** option.

Or, you may see the instructions abbreviated in the following format:

Menu options abbreviated:

Insert | Diagram

Either method can be used, the key is to determine which method to use and use that method consistently. Also, if the abbreviated method is selected, the method of abbreviation should be explained in an introductory section of the manual. For example:

The menu selections will be written in the form of: **menu1 |menu2|menu3**. This shorthand method indicates the sequence of menu selections that will take you to the desired endpoint.

Notice that in each of the previous examples, the menu options were in bold. The font style for emphasis for menu or button options is designated as part of the style guide. Applying bold to text draws a reader's eye to that text and helps the reader easily detect the key words as they move through a series of instructions. The bolded menu name reiterates the vocabulary for the novice and helps to easily identify key steps for the more experienced user. Of course, this assumes that there is a discreet use of bold text. As stated earlier, excess use of bolded words will cause the bolded text to lose its emphasis effectiveness.

When menu options or button names are bolded, you will want to be sure that you avoid the use of bolded text in other normal text. When bold is used for purposes beyond menu or button instructions, the document becomes distracting, making it difficult to read or skim through the instructions.

Guidelines for setting fonts styles are below:

1. Chapter Headings use the largest font.
   a. Each progressively lower level is 1 – 2 points smaller than the above level.
   b. If a font selection is not bold on a higher heading level, it would be not be appropriate to bold the font on a lower heading level.

2. Keep the font type (Arial, Times New Roman, Calibri, etc.) consistent. Change only size, bold or italic.

3. The order of emphasis is as follows:
   a. Size of font creates the most significant impact.
   b. Bold creates the second most significant impact.
   c. Italic creates the third most significant impact.
   d. Combining bold and italic increases the impact beyond bold by itself.

4. Heading styles are only for the use of introducing topics or sections. They will create problems if used in the middle of normal text.

5. Capitalized letters and bolded words can come across like yelling – use with caution.

6. Italicized words can be more difficult to read – the more words in a sequence that is italicized; the harder the text is to read.

Managing Graphics

Managing graphics is significant for User Manuals due to the large number of screen capture and tables that are generally used to describe processes and procedures. Graphics include screen captures, tables, and other non-text objects. The attributes and properties of the screen captures to be addressed in the template or style guide include the outline, size and location or placement.

Frequently, the practice is to frame the screen capture with an outline. When screen captures do not have a clear edge or have a pale background, the image appears to be disconnected and floating. The added outline for the graphic helps "ground" the image resulting in an improved appearance. If outlines are going to be used, they should be used consistently with the parameters established at the beginning of the project.

The parameters for graphic outlines are:

- Outline color.
- Weight of line.

To keep frames softer, use medium gray tones with 1 – 2pixel line weights. Also, when all screen shots have a similar background color, a border with a slightly darker shade of the background color can be used. Using a color coordinated border color will "ground" the graphic while allowing the border appear that it really is part of the graphic. If a black frame is selected, a thinner line weight is required.

Necessary or appropriate size of the graphic can be thought of in the same terms as the size of text. Large, bold graphics with heavy, dark borders draws attention in a very loud fashion – you can definitely see the screen capture or graphic, but you will have a hard time seeing it as part of the flow of the process. On the other hand, if the graphic is too small to see, it becomes pointless. Like text, we want the graphic to be small enough to be able to see the multiple steps and perceive the flow of the process, but also large enough to gather information from the graphic.

When creating screen shots, limit the included content to focus on the items discussed. When the menu bar is the point of the discussion, the screen shot should include only the menu bar and not the entire screen. This will help keep the size of the graphic smaller, while displaying the key points in a large enough to be clearly visible. The only time the entire screen should be displayed in a screen shot is in the introductory screen shot where various parts or components are identified.

While some variance in the position of graphics can help the appearance and flow of the document, there are a few general guidelines to follow. When it is necessary to have a large or wide screen capture, the resulting graphic is placed directly under the text, with the same indents as the text. Notice that this means that the graphic is not wider than the normal text margins. Mid-size graphics generally look best centered under the text. Small graphics (less than half the width of the text) may be placed next to the text, allowing the text to flow to one side of the graphic. Often small graphics are placed to the outside and text to the inside – this assists the reader in quickly identifying an object. For all 3 sizes of graphics, the graphic should be contained within the normal text margins. This ensures that the headings and left margin stay fully visible.

An exception to the placement of the graphics is the introductory screenshot that may be included in the introduction of a chapter. The introductory screen shot is often a graphic of the full screen representing the software to be used in this chapter. The purpose is to call-out all the important features of the software and establish a common terminology which will be carried throughout the chapter or document. For the introductory screenshot, you may choose to use the full page width, so that the details of the screen and the call-outs are readily identifiable. Because the introductory screen-shot is not in the middle of an instruction set, the size does not interfere with the flow of the instructions.

Introductory screen-shots often require call-outs to identify items of importance. While it is possible to create the final graphic with call-outs in the document, it can create an unstable graphic while editing. To create a stable graphic, the screen capture is placed in a separate document or program for editing and processing. After the graphic is complete, a screen-capture is created with all call-outs in place and the file is saved with any supporting graphics for future editing.

Using caption numbering and naming for screen captures and tables allows the writer to use a cross-reference to a specific item. In addition, if there is going to be a Table of Figures at the end of the document, using correct and consistent caption styles will smooth out the process.

Consequently, the standard practice to be applied for caption numbering and naming also needs to be predetermined. For the benefit of the readers, consistency of the location of the caption helps the readers identify the diagram, screen capture, or table.

The factors to predetermine for caption are:

- Caption numbering and naming.
- Table numbering and naming.
- Location of the caption (top or bottom

Guidelines for Graphics

1. Outline
   a. Select outline color
   b. Weight of line

2. Location of graphic
   a. Small –place next to text
   b. Medium – center beneath text
   c. Large – try to keep within the text margins

3. Call-out styles
   a. Color
   b. Shape – square or round
   c. Numbering callouts vs. descriptive text

4. Captions
   a. Location
      i. top/bottom
      ii. centered or left
   b. Determine Use
      i. Table of Contents
      ii. Table of Figures

Use of Lists

The core of the User Manual is the instruction sets and lists are the easiest and the most used method for providing step-by-step instructions. Lists foster quality instructions which help to make procedures easy to learn and are the primary reason someone purchases a user manual. When you want someone to use any given process, providing quality instruction makes all the difference.

When we develop a list, we are trying to identify specific items or steps. The sentence structure for a list of steps in an instruction set becomes very basic with minimal description and no justification or explanation. The introductory paragraph is the place for the descriptive discussion; where it is appropriate to describe the concept and develop the explanation or justification. When writing instructional steps; the words are specific without creative description. While a paragraph is explanatory, a list (and an instruction set) is as brief and concise as possible.

Within the group of lists, we can either have a bulleted list or a numbered list. Bulleted lists are called unordered lists. Numbered lists are called ordered lists. Whether a list is ordered or unordered is related to the dependence of the sequence of the list items. When the sequence of the list items will impact the outcome, use an ordered or numbered list. This means that all instructions are numbered lists because we must do step one before step two. When sequence of list items is not significant, use an unordered or a bulleted list.

In the following example, you can see an application of bullets integrated into numbering:

---

1) Step one
2) Step two

   - For Events 1 & 3, go to step three
   - For Events 2 & 4 , go to step four

3) Step three
4) Step four

As the steps are sequential, they are numbered. The reader knows to go to the following step without specific instruction. In the Step Two process, it does not matter which event is read or processed first; either will result in the same decision or outcome. The events listed as bulleted items under step two are an unordered list.

When creating the list of instructions, keep in mind that the fewer words used, the clearer the instruction will be. Provide a user with a graphic or screen capture when it is helpful to display where to click or to demonstrate the results of their action. When additional explanation is essential to guide the reader from one step to another, the explanation is placed in an informational box or added as an indented paragraph below the instruction.

1) Step One

    The results of step one will automatically generate ABC. You will need to wait for five minutes before proceeding to step two.

2) Step Two

Notice that the additional information for a given step is indented, allowing the sequence from Step One (the previous step) to Step Two (the next step) to be clearly visible. Also, the spacing within the additional information is single-spaced. It is important that the information included within the instruction set is limited to information that will help the reader understand the process or delays that they may experience. The only comments and notes that will appear in the middle of the instruction set are there to assure the reader that they are on track. All warnings and suggestions for the general process should be placed in the paragraph preceding the instruction or set off by placing the information in a text box, a **NOTE:** message or other system.

In general, when there are 2 or more steps to a process, creating a list helps the information stand out. However, we also want to limit the number of steps in a process. Seven steps for any given instruction set is a good target number. While there is not an absolute rule on limit for the number of steps, when there are 12 steps, it may be time to re-evaluate and break the procedure down into smaller processes. Remember, our purpose in numbering steps is to create easy-to-follow instructions.

The purpose for a list is to allow someone to read through the items while they are trying to implement a set of instructions. Keeping the sentences brief with little more than a subject and a verb will help keep the instructions simple. Even though the writer may have some exciting adjectives or adverbs to make the writing appear more creative, they need to avoid the use of unnecessary words. Simple sentences contribute to simple instructions.

Clear instructions are not only simple sentences, but the sentences also clearly identify the noun and/or the subject. That means, avoid using the word "it" in place of a noun. If you are referring to the screen, say, "the screen/ window/ button/ menu will look like…" avoid saying, "it will look like". The writer may think that the word "it" makes sense, but the reader or user, may have their own interpretation. As part of the writing process, we want to remove any possibility that someone will be confused or misunderstand. So, even though it may appear that we have referred to the same noun many times, and the text is not as conversational as the writer may want, avoid pronouns.

Guidelines using lists

1. Create a list there are 2 or more items.
   a. Number the list if the order is important.
   b. Use bullets if the order is not significant.
2. Keep the list to less than 12 items.
3. Avoid explanations within the numbered instruction set.
   a. Explanations go in the paragraph preceding the instruction.
   b. Warnings go in a "Warning" informational text box.
   c. Suggestions and advice goes in a NOTE: text area or box.
   d. When necessary to include explanations; indent explanations and single space.
4. Keep sentences short.
5. Avoid unnecessary words.
6. Use the noun repeatedly; avoid vague pronouns (i.e. it, its).

Index and Glossary

If you know that the project includes an Index and Glossary, it is wise to develop the word list as you are working on the project. At the end of the project, the index will be created from a table of words. As the document is developed, the table of words can be co-created so by the time the document is in editing phases, there is already a list of necessary words to be included in the Index or glossary. The table of words will become a "concordance file" that will be used to create the index.

The index will become a project in itself after the writing is in the final stages. If the indexer has a list to work from, the process is greatly simplified. Also, providing a project manager a list of terms when the document is going through reviews, assures that necessary words have been identified and captured.

The Project Manager may decide to select words from the Word List that need further definition. From this list, a glossary can be created that is added to chapters, sections, or a final glossary at the back of the manual.

Conclusion

Key Benefits of Planning

- Consistency throughout the document.
  - Terminology
  - Formatting
  - Reading level
  - Instruction design
- Less time needed for editing, rewrite, and review.
- Better project management for Project Managers
- Better time management for Subject Matter Experts.

There are no absolutes in creating user manuals and for every guideline established, there will certainly be exceptions. The most important thing to take away from this manual is the establishment of a planning process. At a minimum, you want to establish your outline and assign Heading Levels to the various outline levels before anyone starts writing. It is relatively easy to change the Normal and Heading type and styles if the correct Heading level has been assigned and the remaining text is set as "Normal".

As a result of careful planning process, the document development will flow smoother with the writers focused on writing. Subject Matter Experts will be in demand during the planning process, but will find that once the document outline has been developed, the writers have become quite familiar with the software and are more ready to work with minimal assistance from the Subject Matter Experts. Project Managers will experience a similar involvement; the need for Project Management intervention may be high during the planning phase, but once the execution or the writing begins, they will need less involvement. For both Subject Matter Experts and Project Managers, this will help them manage their own schedules and the Review processes.

www.ingramcontent.com/pod-product-compliance
Lightning Source LLC
Chambersburg PA
CBHW031331290526
45784CB00014B/2552